USA Guide to Paranormal Investigating
(aka Ghost Hunting 101)

"How often have I said to you that when you have eliminated the possible, whatever remains, however improbable, must be the truth?" ~ Sherlock Holmes

So you're interested in the paranormal and learning about hauntings, ghosts, EVPs, and the like. Maybe you've enjoyed watching the latest ghost hunting shows on television or seeing the movies that are out surrounding the paranormal. Maybe you've had an experience of your own that you question or that startled you that has you wanting to learn more and find out exactly what is behind that experience. This guide will help you on your quest to learn more about paranormal investigating and might even teach you about things such as photography, the weather, and using common sense along with your gut instinct.

Ready to get started? Let's go!

What is "ghost hunting" or paranormal investigating? Ghost hunting, or paranormal investigating, is the process of investigating locations that are reported to be haunted, or have paranormal activity. Typically an investigation will consist of a team of individuals working together to collect evidence of paranormal activity. These individuals will employ the use of several types of tools and equipment to scientifically measure and record findings. Such tools include EMF meters, digital thermometers, infrared cameras, thermographic cameras, night vision cameras, handheld video recorders, digital audio recorders, and other types of tools such as flashlights, motion detector lights, dowsing rods, pendulums, ghost boxes and various other electronic devices.

Paranormal investigating has several facets to it and certain procedures you should follow. This handout merely covers just the basics – enough to get you started and have you understand how to properly conduct a paranormal investigation. There are further steps you can take for your investigation once you get these basics down such as completing forms that record the temperature and weather outside, the exact location of your site, wind direction, etc. but for someone just starting out it is more important to get the basics down correctly then add on to your procedures from there the additional steps.

A lot of the information you will read here is mainly just plain common sense but you would be surprised at how many of us can forget basic things in the midst of an investigation.

Make a list of the equipment you plan on bringing with you for your investigation. You might even plan these items out based on the location you are going to investigate - if it is outdoors, indoors, etc. Make sure all your equipment is in good working order, fully charged or has fresh batteries. Also remember to stash additional batteries in your backpack, case or bag just in case you need them later. Battery draining is pretty common when investigating. You might also ensure that you have lens cleaning cloth for your camera or camcorder lenses. Some locations can be dusty or if you are outside you can be facing some dirt filled areas - these can cause potential anomalies which might cause some confusion when reviewing your evidence later. It is a good idea to label your equipment with a permanent marker or label marker so that in the event it gets left somewhere it can be returned to you.

Make sure your cell phone is fully charged. While it is not ideal to carry your cell phone on your person while investigating as it can distrupt other pieces of equipment and cause false positives, it is ideal to have in case of an emergency.

Consider your clothing for the investigation - where are you going to be? Inside? Outside? If you are outside, is it cold out? rainy out? warm out? Check the weather for your location and plan accordingly. If it is cold out consider dressing in layers to stay warm but avoide wearing any materials that might rub and cause noise. You might consider packing hand warmers, foot warmers or even a hat. If you will be outside will you need bug spray, repellant or boots? Dress for your location. You might consider wearing something with a lot of pockets to help hold items or maybe wear a hunter's vest. Rubber soled shoes are ideal for almost any location - you will get a decent grip on surfaces and your feet as well as your toes are protected. If you are investigating an abandoned building, ensure your feet are properly protected as there may be a lot of debris in the building. The material your clothing is made of is another item to take into consideration. You would want to wear a wind breaker that would interfere with your EVP sessions as you are walking along. The same goes for jewelry and keys - they could clink together or dangle making noise that could potentially contaminate your EVP session. And let's not forget perfume and cologne...some spirits like to emit fragrances and even odors to make their presence known. Don't ruin this experience by wearing a scent that could be confused with an actual spirit.

Other items to consider...medication that needs to be taken in the evening - you should already be carrying a first aid kit with additional medications that may be needed in an emergency. Will you need to bring a chair if you will be outside or in an abandoned building? Think about bringing a cooler full of water and drinks to stay properly hydrated as well as some snacks - many of us like to drink an energy drink when investigating and snack on something while manning home base with the DVR system! Make sure your first aid kit has hand sanitizer in it as well as different sized bandages, antibiotic ointment, pain reliever, dust masks and even gauze. You might consider preparing an essentials bag that has toilet paper, tissues, paper towels, your first aid kit, and other items you might need for any investigation location.

Be sure whatever research notes you have taken on your location are packed with your equipment for reference once you get on site. Don't rely on memory to provide you with much needed facts later. Make note of spirits that have been noted at the location but also be open minded to encountering additional spirits that may show up as they are everywhere and some like to get around, often pausing at some places for a while.

Finally, consider every worst case scenario that could possibly happen and be prepared for it. Does your vehicle have a good spare tire and jack in the event a flat tire happens? Do you have jumper cables in case your car battery goes out? Is your vehicle in road trip shape? Do you have plenty of gas? Is your insurance card, GPS system/directions, phone charger and toll tag in the car? If your road trip consists of being on the road for several hours you might consider checking into a reasonable hotel to get some rest prior to driving back. Do you everything you need to get by until you return home - credit cards, toiletries, clothing, etc.? Additionally, if you become ill or need to be hospitalized and unable to communicate how would anyone know who to contact? Make sure you have a list of phone numbers with names to call in the event of an emergency along with a list of medications you take which should be kept in your wallet along with your health insurance card or a copy of it. You might also consider entering an ICE or "In Case of Emergency" number in your cell phone.

THE BASICS OF THE HUNT

1. **Get permission.** Before stepping foot onto any property be sure you have explicit permission from the owner to conduct your investigation. You might even get something in writing in addition to having to sign a waiver form.
2. **Never go on an investigation alone.** Things could get dangerous on a physical and spiritual level so be sure you have at least one partner with you at all times. Carry a cell phone with you but be sure it is on vibrate during your investigation.
3. **Do your homework.** Research your location prior to your investigation for the history of not just the current but the past. This includes interviewing people who have had experiences at the location. Be prepared to debunk.
4. **Ensure your equipment is ready to go.** This means fresh batteries, extra batteries, cleaned lenses, packing done in a manner to prevent damage and free of debris. Ensure you have fresh recording supplies for audio and video recordings if your devices require memory cards or tape. You don't have to have thousands of dollars' worth of equipment to investigate. The basics would include a flashlight, a digital voice recorder, a handheld digital video recorder, a digital camera, and a basic sound enhancement computer program for reviewing EVPs.
5. **Be positive.** Promote a positive energy and be respectful of the deceased. In doing so you are more likely to get a response from a spirit sine they are not threatened.
6. **Conduct a walkthrough prior to investigating.** Take plenty of pictures for comparison later when you are reviewing your findings. Focus more on the area where hauntings have been reported.
7. **Smoking and alcohol don't mix with investigating.** Cigarette smoke is the main cause of a false positive and drinking during an investigation just ruins your creditability.
8. **Log it, write it down.** Make sure you make notes during your investigation as to what you see, hear or feel. In addition, notate whenever you may have contaminated evidence with a

simple cough or sneeze so that you can debunk those sounds later during evidence review.

9. **Never wear perfume, cologne or a distinct scent on your investigation.** Generally a spirit will let you know that they are there simply by emitting some type of fragrance. Friendly entities will emit a pleasant fragrance while unfriendly ones will emit foul odors.

10. **Never form a conclusion on-site.** You want to closely review your findings prior to reaching a conclusion and you may need to do a follow up investigation to answer questions you may come across while reviewing your findings.

ARRIVAL ON LOCATION

You arrive on-site excited and ready to go; what do you do first? Take a deep breath, relax and focus. You need to gather your things, decide what you need immediately and what you can come back for later, or take to a safe room if one is available, and what you will want on hand while investigating.

Most locations, including private residences, you will have someone greet you. You will receive information on where to stay out of, where you can use electrical outlets, where you can set your stuff, etc. Generally you can expect to receive a tour of the location once you get there and get settled with your things. Don't forget to ask questions during the tour as well as take your walk-through pictures for comparison when you get home to do review. Some locations may want you to sign a liability waiver as well. Remember, be respectful and listen to what you are told.

If you are working with a team on-site you may have already decided how the team will be split up for the investigation. If not, after the tour and walk-through would be a good time to talk about how you are going to approach the evening's investigation. Each team leader should have a radio in the event of an emergency, or at least a cell phone with the contact numbers for the other leaders. This might also be a good time to pull out your recorders and go around having each team member say their name in a regular tone of voice followed by a whisper so you are ready ahead of time with tagging of audio files for review later.

Prior to starting your investigation we suggest you say your prayers of protection. No one should be forced to join in but everyone should be respectful of those participating in the team prayer.

INVESTIGATING

Investigations take time, patience and a lot of work. You may not see or hear anything at first when you begin investigating but that doesn't mean when you go back through your pictures and listen to your audio recordings that you won't find anything! Ghost hunting is nothing like what you see on TV shows - you don't always get continuous activity so you have to be patient.

If your intent in participating in a paranormal investigation is to have a ghost chase you or to get scared or startled, you might want to consider staying home and renting a horror flick. Most individuals on an investigation are dedicated to the research and investigating that goes into the investigation itself. Your getting easily scared or startled, running and screaming or freaking out can be a major distraction to those trying to seriously research for paranormal activity.

Focus on the investigation and always try to debunk first before jumping to a false conclusion. If something should happen during the investigation just remember, you have teammates with you. Take a deep breath and begin debunking the activity. If it was a sound could it have been the wind, another teammate or the house settling? If you felt something could it have been your clothing getting caught on something, a teammate brushing by you or your camera strap falling down?

Once you have ruled out all outside influence, THEN begin your techniques to communicate and see if you get a response. Cameras should be ready to take pictures, camcorders should be rolling and you should have already turned on your digital voice recorder. Begin asking questions "Did you throw that object? Could you do it again?" "Did you just knock? Could you knock like me?" (Knock on a wall, door or piece of furniture so that the spirit can imitate you.) Speak in a normal voice, if somewhat quiet voice when doing your EVP work.

Be aware of your surroundings. What floor of the building or house are you on? Where is the nearest exit? Do you feel any drafts from windows, doors or hallways? What normal activities take place at the location? Where are your teammates? Remember - NEVER go off by yourself!

Provoking spirits is NEVER recommended! Try to do or say something normal that might trigger a reaction. If you are in a church, stand up at the pulpit and read a verse from the Bible. If you're in a hospital, pretend to be a patient needing help or open drawers and/or cabinets pretending like you are looking for something in particular asking for help to find it - a bandage, gauze, etc. If you hear music playing and you recognize the tune, try softly singing a little bit of it, just doing that may draw attention and bring a spirit closer to interact with you.

If you have a medium or sensitive with you don't depend on them to lead you straight to the action. They all work on different frequencies and while they may pick up something in one area they may not necessarily pick up everything on the location. If you have more than one with you, they may pick up on different things on the location where they may not necessarily match up. This is because they function on and pick up on different frequencies.

EVP (ELECTRONIC VOICE PHENOMENON) SESSIONS

EVPs are recordings of voice or voice-like sounds that are not audible to the human ear. The frequencies of these sounds are reportedly well below the range of sounds that can be perceived by the human ear. Typically, but not always, they are short having a length of only one word or a short phrase. You will want to use a digital voice recorder for EVP sessions and it is even recommended to have two or three recorders – one that you keep on you and begin recording on from the time you arrive until the time you leave; one that you use for EVP burst sessions; and one that you use during EVP sessions as a stationery recording in the room.

The additional recorders allow you to see if you capture anything different on one than the other. It is also a good idea for review when you hear something on one recorder to check for it on the other recorders.

Let's look at how to capture EVPs and then we'll look at how they are classified.

Capturing EVPs

Below are some basic tips for capturing EVPs:

- Have extra batteries on hand for your voice recorder.
- Speak clearly, never whisper. If someone is whispering make sure and audibly notate that on your recording.
- Keep your recorder away from your mouth.
- Turn off your cell phone or at least set it to vibrate during your EVP session.
- Keep cell phones away from equipment as it could throw distortions on cameras, tape recorders and even EMF meters.
- When you begin your EVP session, catalog as much information on your recording as possible:
  - State the time
  - State the location (what floor, room number, etc.)
  - How many people are with you conducting the session
  - Each person should state their name to have a reference later to their voices
- Be as specific as possible when providing information.
- Do not let anything hit the recorder such as a lens cap or other piece of equipment.
- Talking during the session is okay as long as you do not talk over other people so that EVPs can be clearly distinguished later.
- Audibly note any sound heard as well as notating coughs, sneezes, sniffles, shuffling of feet, etc.
- Do not wear or carry noisy items such as loose jewelry, pocket change, or keys.
- Record in short session to make review easier.
- Ask simple questions and allow at least 20 seconds for an answer. Spirits have

to draw energy to communicate with us so it may take a bit for them to answer.
- Remember, there are spirits all around us so don't get too attached to one spirit that you forget to address the others.

Sample Questions to Ask

Here are some ideas of questions you may want to ask during your EVP session. Be creative and focus your questions around what you learned when doing your research of the location.

- Is there anyone here who would like to answer some questions?
- Are you male or female?
- What is your name?
- Are you married?
- Can you make a noise for me/us?
- Can we take a picture of you?
- Do you want us to leave?
- Do you know you're dead?
- What year is it?
- Why do you like it here?
- Do you want to leave this place?
- Do you want me/us to pray for you?
- How did you die?
- Would you like for us to pass along a message to someone?

Remember, friendly spirits are more willing to communicate when you are giving off positive energy. Experiment with taking turns asking questions – especially if there are members of the opposite sex on your team. Sometimes spirits will respond to one gender over another. Never yell or threaten a spirit – even if they are friendly, they could become angry and pose harm. Ask simple, short questions; less is more. Ask three or four questions then leave...if they have more to say they will let you know.

A few more notes, if there is an occurrence such as someone bumping furniture or a car pulling up outside, be sure to vocally mark it by saying "Note: We heard a loud thump in the hallway."

You may think you will remember these things later but you would be surprised what you might forget.

Classifications of EVPs

There are several different classifications of EVPs. They basically fall into four categories: Class A, Class B, Class C and Class R.

Class A
For an EVP to receive this rating it must be a very clear voice and everyone who listens to it agrees on what is being said without being told by someone else what it says. These voices do not need to be amplified or cleared up using a computer sound editing program but can be clearly heard through the device. It does not have to be loud, just clear about what is being said. The Class A EVPs are the best voices of the dead captured and the rarest to record.

Class B
An EVP that can be understood and most people agree on what is being said will be classified as a Class B. These might not be understood by everyone who listens to it and it might even sound like it is saying something completely different to some people. This class of EVP might need to be amplified using a computer sound editing program before they can be clearly understood. For this rating the voice must be fairly clear and easy to determine what most of the words are when being analyzed by a computer. This is the most common class of EVP to capture.

Class C
A Class C EVP will be the worst quality voice that you can capture. It is nearly impossible to understand what is being said even with the assistance of a computer program. These are often just whispers or mumbled words that might even sound robotic. The voice cannot be understood but you still know it's an EVP because of the fact that no one was talking during the recording session and human sounding voices can be clearly heard in the background.

Class R
A Class R rating means the EVP will have a meaning when played in reverse. Some will have a meaning when played normally and a different meaning in reverse. When this happens the EVP will receive two classifications. For example, a Class A EVP with excellent and clear meaning in reverse as well will receive a Class A-RA rating. Meaning it was very clear to understand both forward and in reverse. It can have a Class A-RC rating because that would mean that it could not be understood in reverse causing it not to be a Class R EVP. You may have a Class B-RB or a Class A-RB.

TAKING PICTURES

More experienced photographers can probably give us a better idea of the technicalities of cameras, flashes and the like so we won't address those topics here. If you want to learn more about these types of topics we would suggest contacting a local photographer in your area who is not associated with the paranormal or reaching out to our sponsor, Scott Frederick with Scott Frederick Photography. There are also additional pointers regarding avoiding false positives on the National Paranormal Society website under the Photography tab.

As we mentioned initially in the Arriving On-Site section, be sure and take pictures of the location prior to beginning your investigation. These pictures will be vital to your review process once you get home so that you can compare back to them when you think you may have caught something. This is an important step in debunking.

When an investigator is documenting with pictures, always remember the 1, 2, 3 rule - take three pictures in row when you are taking pictures of the location. This gives you a chance of showing movement as well as a sudden appearance or even disappearance. Be sure you watch where your fingers are on the camera - keep them away from the lens itself as well as preventing them from blocking your flash. You would be surprised how many people think they have caught a shadow figure in a picture only to realize it was the tip of their finger blocking the flash. Be aware of what you are wearing - caps and bulky clothing can sometimes cast shadows. Tuck away any camera cords, neck straps or the like to prevent them from swaying or swinging into the picture. Don't worry about reviewing the pictures as you go unless you saw something with your own eyes; otherwise you might lose the opportunity to snap activity around you.

A few tips to keep in mind while you are taking pictures and shooting video:

- Avoid taking pictures or video during any kind of weather conditions or where smoke is visible. This can contaminate your findings.
- Hold your breath when taking pictures on a cold night.
- Hold the camera still as movement will distort the photo.
- Never take photos of shiny or reflective items.
- Learn and know what your equipment can and cannot do. Test it out before going on an investigation.
- Never take a photo or video towards a light. It will create a glare and destroy any possible evidence you may or may not have.

THINKING OUTSIDE THE BOX

In addition to your typical equipment and techniques, there are a few tricks to keep in mind as well.

A dollar bill or tape measure can be handy when it comes to referencing the size of something or for movement. You could place something like a ball at the end of the tape measure, ask the spirit to move the ball, leave to conduct your research in another area of the location then come back and see if the ball has moved.

Baby powder, or even flour, can be a great trick for catching footsteps. Make sure you bring a tarp or black plastic table cloth to use for this so that you don't mess up the client's location. Sprinkle it down on the tarp or table cloth and throughout the night return to see if it has been disturbed. If you do capture something, use the dollar bill beside it then take a picture to give an idea of the size.

Flashlights are handy little tricks of the trade as well. Used not only for safety and finding your way around in the dark, they can be handy communication devices by turning the off and asking a spirit to communicate through it by turning it off or on. Be sure that the flashlight you use does not turn on easily by vibrations of such things like the air conditioning unit. Sometimes you might want to use more than one flashlight so always have a few on hand with fresh batteries as well as extra batteries nearby in case you need them.

Trigger objects are wonderful tricks of the trade. Items such as toys, balls, balloons, tools, etc. make great trigger object to elicit responses. Children spirits are drawn to such items and you might even get them to interact more when using something that they are attracted to.

Dowsing rods can be tricky to use as the most important part in using them is establishing what positions will be used for neutral, yes and no. The Crew uses the rods pointing straight forward as neutral, crossing as yes and pointing away as no. Think of a control question to determine whether you are dealing with an intelligent spirit or not. Something like asking if your middle name is John or something similar to determine if the spirit answers correctly. Keep your hands steady as your conduct your dowsing rod session. You may find some playful spirits would like to twirl the rods round and round - let them have fun spinning the rods as you capture it on video.

HAUNTINGS

A haunting is a recurring presence of a ghost, demon, or similar supernatural being at a specific place. A haunting can occur anywhere. Belief in hauntings and ghost is worldwide and recorded throughout history. Anthropological evidence indicates it was common in prehistoric times. Typical researchers of hauntings consist of parapsychologists, paranormal investigators, historians, folklorists, anthropologists, and of course, skeptics. Let's take a look at the different types of hauntings.

Intelligent
The entity is aware of its surroundings, including living people who are present within the area. The entity's classification can be benevolent, benign or malevolent.

Residual
It is believed that a residual haunting occurs when an entity visits the same place and performs a repetitive act. The entity is not cognizant of any living persons who are present within the area and often times is not aware of its surroundings as we see it today. These types of hauntings typically occur after a tragic event and the repetitive acts display a portion of the event as it happened. Sometimes the residual act is of a mundane act the individual repeated often in life. Generally these entities are not considered an actual ghost but instead an imprint of energy left at the location due to the tragic event.

Benevolent
This haunting is of a protective nature. It is felt that the entity is a loved one looking to protect the living people within an area in which a demon or other possibly violent entity exists. These ghosts are often found to be related to the living persons within the area but have also been found to be closely connected to those being attacked.

Malevolent
This haunting occurs when a ghost or demon seeks to inflict harm on the living within an area. It is believed that this occurs because the entity is angry about events that occurred in its life, is jealous of the living, or is a malicious personality in general who is seeking attention or being defensive of its home and wishes current living persons to leave, or is sad and wants others to acknowledge its sadness and feel its misery. Poltergeist activity is often confused with a malevolent haunting; usually it is a spirit seeking attention. A demon can be present if a murder occurred in the area. These types of hauntings are rare but do exist. We will briefly go into demonology – just enough to give you a basic understanding of them.

Benign
This type of haunting occurs when the entity is unconcerned with the living or unaware of their presence in the area. This type can also be connected to an intelligent or residual haunting.

If you choose to say prayers before and after your investigation, here are a few of our favorites we like to use. All prayers should be said in a circle with all team members holding hands. Concentration on the words of the prayers is important so that the prayer maintains its power.

PRAYERS FOR BEFORE THE INVESTIGATION

St. Michael's Prayer
Saint Michael the archangel, defend us in battle. Be our protection against the wickedness and snares of the devil. May God rebuke him, we humbly pray. And do thou, O prince of the heavenly host, by the power of God, thrust into hell Satan and all evil spirits who wander through the world for the ruin of souls. Amen.

Calm Protection Prayer Prior
Heavenly Father, we gather here tonight as investigators to better understand the universe you have given us. We ask you to watch over us, to guide us and keep our minds focused and free from fear. We ask for your protection from any that would do us harm and hope you will give understanding to those we seek, assuring them of our intent and that they have nothing to fear from us as investigators. Again we ask you to see us safely through this night. Amen

Strong Protection Prayer Prior
Our Heavenly Father, we come to this location tonight to better understand the unknown. We do, as investigators, know that there are dangers in searching for these truths. We come

to you and ask you to protect our hearts, minds, bodies, and souls from any and all inhuman, malevolent, foul, or evil spirits. We call on your Arch Angel Michael to stand by our sides. We ask your guidance, so we may get through this night safely without harm to your children on either side. We once again ask that you protect us from all harm and danger. Amen.

Prayer Before the Hunt
May we be strong in the Lord and in the strength of his Might! St. Michael, the archangel, shall defund us and be our protection against all evil and negativity on all levels. May the Divine light of the Archangels and all the Ascended masters surround us with their love. We call on Metatron and ask that all negative thought forms, lost souls, negative residues, negative elementals, and fragments be permanently healed and taken into the light, that all may be freed according to the highest will of God. May we be purified and blessed upon every level of our being, in the work we do, and be given more Divine power and protection. Thy will be done! Thank you and Amen.

PRAYERS FOR AFTER THE INVESTIGATION

Closing Prayer
In the name of Jesus Christ, I command all human spirits to be bound to the confines of the cemetery. I command all inhuman spirits to go where Jesus Christ tells you to go, for it is he who commands you. Amen.

Calm Protection Prayer After
Lord, thank you for seeing us safely through this investigation. We brought no spirits with us and ask you let none follow us from this location, other than the loved ones we carry with us always. Finally, we ask for your continued protection to see us safely home. Amen.

Strong Protection Prayer After

We are speaking to any and all entities who have chosen to follow us out of this location. We command all human and inhuman spirits whom want to follow us home to stay at this location. You are not allowed to attach to us or our equipment. We bind you to this location and command you to return to where you came from. We do this with the authority Christ gave us. Heavenly Father, we ask you see us safely from this location and protect us on our journey. Amen.

Simple Closing Prayer

In the name of Jesus Christ, I command all human spirits to be bound to the confines of the cemetery. I command all inhuman spirits to go where Jesus Christ tells you to go, for it is he who commands you. Amen.

Closing Prayer

God bless every corner of this house, may peace dwell within. Protect all that come and go, whether friend or kin. Bless every door and window pane, and every ceiling and wall. Bless every closet, nook and cranny, crawl space or basement, bless it all! Bless the roof and ground surrounding with your protective love and light. Hold us in your love care every second of every day, in every way from early morning into sheltered night. Let all be in your complete perfection as you intended. Release all negativity into your confirmed light that is extended. We thank you and expect your miraculous intervention. Clearing all with purification, love, peach and joy as divinely intended. Amen

PROTECTION PRAYERS (to be said after an opening prayer)

Pre-Investigation Prayer of Protection

Heavenly Father, God of the Universe, we all come to you in prayer tonight for our protection. Please protect our bodies, souls, equipment, and all personal possessions from any harm. We ask that you send your angels to guard us against all evil and malevolent spirits and protect us with your white light. We give permission to your angels to intervene for us on our behalf. We pray that you allow the spirits to show themselves to us and allow us to obtain evidence of their existence. Lord, let our endeavors tonight not be in vain. Please allow us to have a safe an eventful investigation tonight. We pray that no spirit or ghost, of evil will or intention, be allowed to attach themselves to any of us or our belongings and that they not be allowed to follow us to our homes or enter therein. We pray Lord that you also protect all of our fellow paranormal investigators from any harm tonight and allow us all a safe journey to our homes. We pray these things in the name of the Father, the Son (Jesus Christ) and the Holy Spirit. Amen.

Simple Protection Prayer

Visit Lord we pray this place, and drive from it all the snares of the enemy, let your holy angels dwell here to keep us in peace, and may your blessing be upon it evermore through Jesus hrist our Lord. Amen.

The Hedge Prayer

Trusting in the promise that whatever we ask the Father in Jesus' name he will do, we now approach you, Father, with the confidence in our Lord's words and in your infinite power and love for us. With the intercession of the blessed Virgin Mary, mother of God, the blessed apostles Peter and Paul, the blessed Archangel Michael, our guardian angels, all the Saints and Angels in Heaven, and holy in the power of his blessed name, we ask you Father to protect us and keep us from the harassment of the devil and his minions. Father, we ask that you build a Hedge of Protection around us and to help keep the hedge repaired and the gates locked so that the devil and his minions have no access or means to breach the hedge except by your expressed will. Father we know that we are powerless against the spiritual forces of evil and recognize our utter dependence on you and your power. Look with mercy upon us. Do not look upon our sins, O Lord; rather look at the sufferings of your beloved Son and see the victim whose bitter passion and death has reconciled us to you. By the victory of the cross, protect us from all evil and rebuke any evil spirits who wish to attack, influence, or breach your Hedge of Protection in any way. Send them back to Hell and fortify your hedge for our protection by the blood of your Son, Jesus. Send your Holy Angels to watch over us and protect us. Father, all of these things we ask in the most holy name of Jesus Christ, your Son. Thank you Father for hearing our prayer. Amen.

Ring of Protection Prayer

In the name of all that is goodness and light, surround our circle in the white light of holy protection. We ask that no harm befalls or follows the protected circle and that our quest benefit all who are among us. In the name of all that is goodness and light, we thank thee for your protection of holy white light.
Amen.

Ground Blessing

Our help is in the name of the Lord who has made heaven and earth. The Lord be will you and with your spirit. Let us pray. Lord God almighty, bless this land. May health, chastity, conquest of sin, virtue, humility, goodness, and meekness flourish here. May the law be observed in its fullness, and thanks be given to God the Father, and the Son, and the Holy Spirit. And may this blessing always remain on this land and on those who live in it, now and forever and ever. Amen.

St. Christopher Prayer for Travelers/Prayer for a Safe Journey

Lord, we humbly ask you to give your Almighty protection to all travelers. Accept our fervent and sincere prayers that through your great power and unfaltering spirit, those who travel may reach their destination safe and sound. Grant your divine guidance and infinite wisdom to all who operate automobiles, trains, planes and boats. Inspire them with due sense of duty and knowledge and help them guide those entrusted in their care to complete their travel safely. We thank you, oh Lord, for your great mercy and unending love to all mankind and for extending your arm of protection to all travelers. Amen.

WRAPPING UP

As with anything, practice allows you to get better with your investigation techniques and methods. Remember to be patient throughout your investigation. A lot of work is put into preparing for your investigation and you shouldn't expect to get something right away. Most of the time you will think you had a non-productive investigation only to get home and hear a few EVPs you caught.

Anomaly – an irregular or unusual event which does not fit the standard rule or law. An anomaly is something that cannot be explained by currently accepted scientific theories. Anything weird, abnormal, strange, odd or difficult to classify is considered an anomaly.

Anthropomorphic – ascribing human form or attributes to a being or thing not human, especially to a deity; resembling or made to resemble a human form.

Apophenia – the spontaneous perception of connections and meaningfulness of unrelated phenomena.

Apparition – appearance of a spirit.

Aura – invisible to the naked eye, it is a glow surrounding an individual and changes color and form depending on the mental and physical well-being of the individual.

Automatic Writing – when a person can produce writing that is not their own writing style and can convey messages from the deceased.

Channeling – the method in which mediums allow themselves to be used in order to manifest something which comes from outside themselves.

Clairvoyance – supernatural power of seeing people, places, things and events outside the space and time of natural viewing.

Cryto-zoology – the branch of paranormal research which deals with the exploration of legendary creatures.

Déjà Vu – seeing or doing something completely new but having the distinct feeling that the experience had been done before.

Demon/Demonic – a supernatural malevolent spirit that causes harm and/or extreme emotional distress.

Demonology – the study of demons

Disembodied Voice – a voice that is heard in real time (unlike an EVP) that is not coming from any corporeal form.

EMF (electro-magnet field) – classically, the electromagnetic field is a physical influence (a field) that permeates through all of space, and which arises from electrically charged objects and describes one of the four fundamental forces of nature – electromagnetism. Ghost activities can sometimes cause changes in the electro-magnetic field and are measured with an EMF meter. There is a theory that a high amount of electromagnetic energy can cause poltergeist activity but also a theory that these same high energy levels attract spirits.

Empath – an individual who is particularly sensitive to the psychic emanations of his or her surroundings, even to a degree of telepathically receiving and experiencing the emotions of others in their proximity. Obviously, psychic empathy can be regarded as a mixed blessing and the empathy must learn to gain a measure of control over this ability.

Entity – an entity is something that has a distinct, separate existence.

ESP (extra sensory perception) – perception that involves awareness of information about something (such as a person or event) not gained through the senses and not deducible from previous experience.

Evil – refers to the morally or ethically objectionable behavior or thought.

EVP (electronic voice phenomena) – the communication by spirits via tape recorders or digital voice recorders. These communications cannot be heard at the time but will be audible when the recordings are played back. They are classified into four classes: Class A, Class B, Class C and Class R.

>   **Class A:** Must be a very clear voice and everyone that you let listen to the recording agrees on what is being said by the ghost or spirit. Everyone who hears the voice must come to the same conclusion about what it is saying without being told by another investigator. These voices do not need to be amplified or cleared up using a computer sound editing program but can be clearly heard straight from the recording device. It does not have to be extremely loud but it must be clear on what is being said. Class A EVPs are the best voices of the dead captured and are the rarest to record.

>   **Class B:** Can be understood and most people agree on what is said. These might not be understood by everyone who listens to them and might even sound like it is saying something completely different to other people that listen to the recording. This class might need to be amplified using a computer sound editing program before they can be clearly understood. To get this rating the voice must be fairly clear and easy to determine what most of the words are when analyzing with a computer. This is the most common class of EVP captured.

>   **Class C:** The worse quality voices that you can capture. It is nearly impossible to understand what is being said even with the help of computer enhancement. These are often just whispers or mumbled words, or might even sound robotic. The voice cannot

be understood but the investigator still knows that it is an EVP because of the fact that no one was talking during the recording session and human sounding voices can clearly be heard in the background noise.

>   **Class R:** Must have a meaning to it when played in reverse. Some EVPS will have a meaning when played normally and a different meaning in reverse. When this happens it will have two classifications. For example, a Class A EVP with an excellent and clear meaning in reverse as well would be titled a Class A-RA, EVP. This means it was very clear to understand both forward and in reverse. It cannot have a Class A-RC because this would mean that it could not be understood in reverse which would not be a Class R EVP. You may have a Class B-RB or a Class A-RB, etc.

Exorcism – the practice of evicting demons or other evil spiritual entities which have possessed a person or object. This is an ancient practice that requires great skill and should not be attempted by anyone except trained clergy members.

False Positive – believing something to be true when in fact it isn't.

Ghost – a non-corporeal manifestation of the spirit or soul of a dead person that has remained on Earth after death.

Ghost Hunting – the act of using scientific instruments to detect and attempt to communicate with spirits.

Guardian Spirit – a spirit who protects and guides a particular person.

Haunting – to inhabit, visit or appear in the form of a ghost or other supernatural being. There are four distinct types of haunting: Residual, Intelligent, Poltergeist, and Demonic.

Residual: an imprint left behind by an event with high energy that may have been extremely emotional. The manifestations seen are not aware of your presence and will repeat their actions time and time again.

Intelligent: a haunting where the entity is conscious and can interact with living witnesses.

Poltergeist: an extremely rare occurrence wherein random objects are moved and sounds produced by an unseen force, the sole purpose being to draw attention to itself. The phenomenon always involves a specific individual, frequently or a child or adolescent.

Demonic: a supernatural malevolent spirit that causes harm and/or extreme emotional distress. Usually associated with foul odors, visible wounds and aversion to sacred objects.

Incorporeal or Non-corporeal – without the nature of a body or substance.

ITC (Instrumental Transcommunication) – any communication with a spirit that occurs in real time with the help of an electronic device such as a radio or television. The spirits appear on a television in the static or can be heard on the radio in between frequencies.

Intelligent Haunt – a haunting where the entity is conscious and can interact with living witnesses.

Intuition – an immediate form of knowledge in which the knower is directly acquainted with the object of knowledge. Intuition differs from all forms of mediated knowledge, which generally involve conceptualizing the object of knowledge by means of rational/analytical thought processes.

Intuitive – a person sensitive to the feelings of other life forms, as well as signals of nature.

Ley Lines – alignments of ancient sites; these are considered to be earth's natural energy lines. Spirits may use these lines as a way of traveling quickly from one place to another. It has also been suggested that where two ley lines cross there is a possible chance of a portal opening to other dimensions.

Malevolent – a malicious supernatural force that causes harmful acts to living beings.

Manifestation – the materialized form of a spirit

Materialization – a ghost appearing visually, suddenly or gradually, sometimes indistinct, sometimes seemingly quite solid.

Matrixing – the natural tendency for the human mind to interpret sensory input, what is perceived visually, audibly or tactilely, as something familiar or more easily understood and accepted; in effect, mentally "filling in the blanks." Refer to Pareidolia.

Medium – a person who possesses the ability to communicate with spirits of deceased people, and occasionally animals.

Metaphysics – the branch of philosophy concerned with explaining the ultimate nature of reality, being, and the world. More recently, the term has also been used more loosely to refer to "subjects that are beyond the physical world."

Near-Death Experience – an experience reported by a person who nearly died, or who experienced clinical death and then was revived.

Orb – name given to typically circular anomalies appearing either in photographs or with the naked eye. They are allegedly the true spirit form and one theory is that these are seen at the beginning of a manifestation. Orbs commonly appearing in photographs or on video surveillance are usually dust, bugs or other airborn particles; however, those seen with the naked eye may not always be one of these explanations.

Out-of-Body Experience (OBE) – an experience that typically involves a sensation of floating outside of one's body and in some cases seeing one's physical body from a place outside one's body.

Pareidolia - the imagined perception of a pattern or meaning where it does not actually exist, as in considering the moon to have human features.

Paranormal – an umbrella term used to describe a wide variety of reported anomalous phenomena that lacks scientific explanation.

Parapsychology – the study of seeming mental awareness of or the influence upon external objects without any physical or energetic means of causation which scientist currently understand. Another definition of parapsychology is the scientific study of paranormal phenomena.

Poltergeist – an extremely rare occurrence wherein random objects are moved and sounds produced by an unseen force, the sole purpose being to draw attention to itself. The phenomenon always involves a specific individual, frequently or a child or adolescent.

Possession – where a person and sometimes an animal becomes overtaken by a being that is not their own.

Portal – a doorway, entrance or gate between spiritual and physical worlds.

Quantum Mechanics – fundamental branch of physics with wide applications in experimental physics and theoretical physics that replaces classical mechanics and classical electromagnetism at the atomic and subatomic levels.

Residual Haunting – an imprint left behind by an event with high energy that may have been extremely emotional. The manifestations seen are not aware of your presence and will repeat their actions time and time again.

Sacred (Holiness) – state of being holy or sacred.

Sensitive – someone who can sense a paranormal presence.

Shadow Person – a being which has never lived on earth and never will. These are often associated with negative energy.

Sleep Paralysis – temporary paralysis of the body shortly after work up or shortly before falling asleep. Described as a heavy feeling that some confuse with possession.

Smudging – the burning of dried herbs, most often white sage, to purify a house from a malevolent spirit.

Spirituality – the act of being spiritual; not necessarily religious.

Supernatural – refers to the forces and phenomena that are not observed in nature and therefore beyond verifiable measurement.

Telepathy – communication of information from one mind to another by means other than the known perceptual senses.

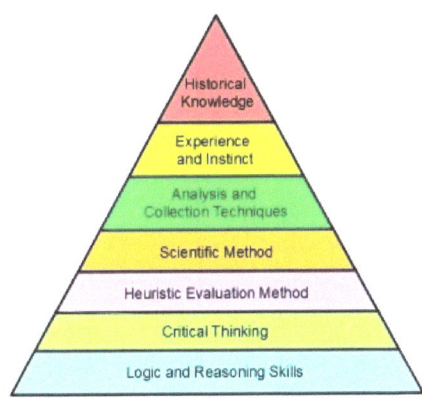

Meet USA Paranormal's new junior investigator. " Monica Busekrus.. Ready to take on the challenges in paranormal research.

Monica feels the passion of what it takes to become a great paranormal investigator. Willing to help those when they have nowhere too turn.

In my quest to become a competent investigator of the paranormal I've spent a lot of time trying to answer this question: What make's a great paranormal investigator? The pyramid on the left represents the primary factors that determine the ability for an investigator to evaluate paranormal claims effectively and objectively.

Factors on the bottom are considered more important or more fundamental than factors higher up the pyramid. Skills higher up the pyramid should be built on a strong foundation. Having a great deal of experience investigating paranormal claims means little if you don't have solid logic and critical thinking skills.

Logic and Reasoning Skills: This includes basic understanding of what logic is but more importantly is the ability to apply logic objectively. I have seen many examples of investigators succumbing to logical fallacies at an amazing rate. No one is perfect, but you should at least no what makes a logical argument, common logical fallacies, and be able to correctly generate logical conclusions the more often than not.

Critical Thinking: Critical thinking is the ability to identify faults, information gaps, and other explanations. Critical thinking is, well, critical to investigation of paranormal claims. An investigator needs to be able to see the holes in a narrative or be able to identify alternative explanations for a phenomenon.

Heuristic Evaluation Method: Heuristics are "rules of thumb" used to evaluate the validity of an explanation. A common heuristic is Occam's Razor. These methods are used to weigh competing explanations, identify information gaps, and ultimately determine which explanations to pursue for scientific testing. A good investigator should have a reasonable heuristic evaluation method picked and be familiar with applying this method to paranormal claims.

Scientific Method: The Scientific method is the next level of testing. Scientific testing can be difficult, expensive and in some cases it may be impossible to design an adequate test. However scientific testing of a hypothesis is the ultimate goal. Investigators should have a solid understanding of the process and requirements of the scientific method.

Analysis and Collection Techniques: These are the techniques used to collect and analyze evidence to determine validity and value. All of these techniques should be based on the standard of the scientific standard, reproducability. This means investigators should establish guidelines for using equipment that includes the careful recording of data and use equipment in a way that avoid false positives. For example, waving a Trifield meter around can skew the results. Any movement can move the needle generating a false positive.

Experience and Instinct: While experience and instinct needs to be limited while analyzing evidence it is extremely important during the collection stage. Experience with investigations and intuition can lead an investigator to find evidence or draw out details from a witness that would have otherwise been missed.

Historical Knowledge: Knowledge of the history of paranormal research and various phenomena can be very useful in investigations. Often this knowledge can fuel intuition or provide insight into alternative explanations.

Secondary Factors

Interpersonal Skills: The ability to communicate effectively with others. This may be more important depending on what the investigation entails. An investigator who interviews witnesses should have strong interpersonal skills while an investigator who analyzes media and reports doesn't require strong interpersonal skills. But being able to communicate effectively is always a plus.

Professionalism: The ability to carry yourself in a professional way. This is influenced by the way you dress, attitudes, habits, and the way you communicate with people. Professionalism is always a desirable attribute for investigators.

Community Awareness: Awareness of what is going on in the paranormal and skeptical communities is valuable. Being "plugged-in" can provide access to valuable resources that you would be otherwise unaware of.

Patience and Persistence: Investigations can be long and tedious. People love closure and investigators especially want to be able to say this was the cause. This is what I think is a driving factor behind a lot of the sloppy conclusions presented by believers and skeptics. You must be patient and persistent with investigations. Let the evidence guide conclusions, not impatience.

Area Expertise: Expertise in certain subject areas can be very important. If you are studying UFOs expertise in astronomy and meteorology would be very helpful. If you analyze media like pictures, video and audio, it is a good idea to start reading up on forensic analysis techniques. If you interview witnesses than an understanding of human psychology is important, specifically what conditions affect a person's ability to make an accurate observation. Other areas include media forensics (photos, video, audio), statistics, astronomy, photography, videography, video and photo editing, meteorology, geology, biology and zoology.

If you are serious about becoming a paranormal investigator please spend some time developing your methods and yourself before grabbing the trifield meter. The field, more than ever needs objective investigators with a toolbox full of analysis techniques.

Paranormal Investigator Certification Course

Third Edition

Excerpts from the Course

The first page of each chapter along with sub points is included below for your review. Many pictures and charts are included within the chapters of the actual Course. Page numbers have been omitted from this excerpt, but appear in the actual Course.

Enroll in the CPI Course Here: http://www.paranexus.org/forum/index.php?page=51

Course is available with or without certification

ParaNexus™
Where Paranormal Researchers Connect

Paranormal Investigator Certification Course

**A Comprehensive Study Program
for Serious Researchers and Investigators**

Third Edition

Doug Kelley, CH, CSL, CFPI

Paranormal Investigator Certification Course

By Doug Kelley CH, CSL, CFPI

Course Code: CPI-SPIRIT3

Continuing Education Units (CEU): 2.5

Published by:

    ParaNexus Publishing, www.ParaNexus.org

ABOUT CERTIFICATION

Upon successful completion of this course, you will receive your Certificate of Completion via email. You will then earn a *Certified Paranormal Investigator* (CPI) designation from the Life Leadership Institute, which certifies that you have completed a comprehensive course on paranormal study. You may then add the "CPI" designation after your name.

ABOUT ACCREDITATION

Currently, no academic institution accredits paranormal investigation courses. Since this course was designed to be a comprehensive and high quality study of the field of Paranormal Investigation, the author has chosen to accredit this course through the Life Leadership Institute, a division of his training corporation, Kelley Training Systems, Inc. Accreditation is in accordance with the guidelines set forth by The International Association for Continuing Education and Training (IACET). Accreditation adds to the credibility, professionalism, and respectability of the student who successfully passes this course. Hopefully more academic institutions will begin to accredit paranormal investigation courses as the field matures and humanity's understanding and acceptance of unexplainable events evolves.

2.5 CEU's are awarded to students who successfully complete this course with a passing score. Please Note: Continuing Education Credit approval is at the discretion of your academic institution or licensing board; not all educational courses are approved by all institutions and boards, which is common in the academic world. An official permanent record of your successful completion of this course is included with your certificate. To replace a lost copy, send your request to doug@dougkelley.com and include your full name, the date in which you completed the course, and the email or regular mail address of the recipient.

Dedication

To all those special and self-aware people who thirst for knowledge
and hunger to understand the grander picture...

To all the people who are not satisfied with the status quo, and who want to
think for themselves in answering the profound questions of life and the Universe...

To those who hunger to fulfill their human nature of exploring the unknown...

I salute you and dedicate this course in your honor.

Acknowledgements

This course would not be nearly as good without the help and input from many others. I wish to thank all of the members of the SPIRITeam for being among the finest Paranormal Investigators in the field. I'm proud of you all and am honored to call you "family."

Specifically, I would like to acknowledge and thank Chris "Q" Quattlebaum for his creativity in researching orb photography including his test pictures; Kristi Robinson for her input on working with Sensitives; Gregory Kent for educating us all on table tipping; Jari Mikkola for allowing me to use mist pictures; Linda Isbell, Grant Rubendunst, and Shauna Isbell for their enthusiastic feedback on specific chapters; Kim Quattlebaum for being an enthusiastic Chapter Director and supporter of all things SPIRITeam; Stan and Simone Smith, Scott Bruneau, Jared Gray, and others for their enthusiastic support of this course.

I would also like to thank my best friend and partner, Tracy Kelley, for her untiring and invaluable help in editing this course, as she has with all of my work. I would not be who I am without you, Sweetie.

Acclaim for this Course

"The best paranormal investigation training course I've taken to date! Doug Kelley, founder of the SPIRITeam and author of this course deserves kudos from the entire paranormal community for taking the time to design this easy to use, inexpensive and interactive training course. Doug's humanistic, logical and sensible approach to teaching is easily understood by any level of student in the field. Whether you're just starting out—or are a veteran—anyone will benefit from taking this course.

"Not only will the student learn how to use the equipment to view and analyze collected evidence, learn terminologies and their proper usage, how to accurately log a case from beginning to end (and what to do with that information), how to find cases to work & ways to land those cases, how to deal with clients on ALL levels, the do's and don'ts of working within a team whether on a case or not, but—it makes the student take a long look inside themselves to see not only why they wish to enter into the field of the paranormal, but if the field will suit them.

"Being a paranormal investigator is much, much more than learning how to use a Gaussmeter—it's about helping people, working well with team members—within your own team, and with others outside your team—and networking in the whole of the paranormal community. Doug Kelley hits an over-the-fence home run with all of these points in his paranormal investigators training course."

—**Dusty Smith, CPI**, President & Founder of The Daytona Beach Paranormal Research Group, Inc., and author of *Dread and the Dead Filled the Dunnam House*.

"**I have spent an unbelievable amount of money on books relating to ghosts**, orbs, you name it. If it dealt with the paranormal, I had it. Your course is amazing in the fact that it covers every detail. When I read the chapter on photography, I was furious due to the fact I took a class on photography and spent hundreds of dollars on it, only to find I learned more in ten pages in the course than I did in weeks of class. I have even bought a course on the paranormal and it was nothing and I mean nothing like what I have found here. The added plus to your course is the fact it is written in a way you understand every chapter. Would I recommend this course? Absolutely, without a doubt. Why spend more money on every new book you can find on the paranormal when you can find everything you need in one course. I also have to say this course is worth far more than what you are asking and I would pay it as I'm sure anyone who spent what I have on all the other texts and courses out there would. I can't say enough about it other than I wish it was out sooner and thank you for taking the time to write what everyone truly needs to know and have it all in one course. Thank you again!"

—**Linda Gunn, CPI**, Founder of CrossLight Paranormal Research, Jacksonville, FL.

"**This comprehensive Study Program is, in my opinion, the best on the market, hands down!** This most certainly isn't your typical "Ghost Hunting" manual…. I got so much more than I expected at the low price I paid. The amount of useful information went way beyond any of the previous courses I've taken and gave me a deeper level of understanding of what a true Paranormal Investigator does. Here you will find a Course packed full of useful information for intelligent, truth-seeking people. The skills you gain as a result of taking the Course have greater implications than first meets the eye. This is a real-world guidebook into the exciting world of Paranormal Investigating. As an educated professional, I highly recommend this Certification Course. You will know without a doubt that you are truly a Paranormal Investigator and have the skills, tools, and education to help others understand the nature of that which is unknown to most."

—**Matt Reynolds, CPI**, Southern Adirondacks, NY

"**This is a wonderful course to use if you are investigating the paranormal.** Doug Kelley is a great instructor, giving you all the knowledge that you need to start investigating. Some of the information within Doug Kelley's textbook is not found in other Paranormal Investigator books, and it is information that all beginning paranormal investigators need. Doug and his team are there all the time to answer any questions you may have, and they do it very quickly. I highly recommend this course to anyone and everyone."

—**Kelly Davis, CPI**, Founder of Macoupin County Ghost Hunters, Bunker Hill, IL.

"**I've been a paranormal investigator for about 6 years,** and up until now have been hesitant, even skeptical about the idea of certification courses. I must say that this course was a very useful and enlightening addition to my personal experience as an investigator, and I recommend it to anyone that might be considering a paranormal certification course."

—**Joey Ward, CPI**, Founder of West Georgia Paranormal Research Society, Woodbury, GA.

Contents

Contents, continued

INTRODUCTION

Welcome to the Paranormal Investigator Certification Course!

I n this course you will learn everything necessary to start you off on the right foot as a Paranormal Investigator. While not designed to be a graduate course in Parapsychology, it is comprehensive and includes important information that you will not find in any other paranormal course or online. It has been written specifically to help SPIRITeam members and others to be among the most professional Paranormal Investigators in the field.

Once completed, you will take the final, open-book exam to receive your Certificate of Completion. You will then earn a *Certified Paranormal Investigator* (CPI) designation from the Life Leadership Institute, which certifies that you have completed a comprehensive course on paranormal study. You may then add the "CPI" designation after your name. Once completed, you will be all set to further your knowledge and experience by performing paranormal investigations!

The field of paranormal research and investigation has gained widespread attention over the past few years. This evolving field is wide open, and there are no "experts," only knowledgeable and wonderful people. And that's what this course is designed to do—enhance your knowledge of the paranormal field in a fairly short time frame.

But this course goes way beyond simply describing the latest electronic gadgets used in paranormal investigations. While it does explain equipment and the basics of operation, this course discusses what it really takes to be a top-notch Paranormal Investigator: an enhanced understanding of the human condition, which then gives clues and insight into the "spirit condition." As you will learn, the differences between humans and spirits are minimal at best.

You will learn how your own beliefs come into play on investigations. You will learn about many of the types of hauntings, and how to resolve them. You will learn about Psychics, Sensitives, and Mediums, and how to work effectively with them. You will learn about spirit communication, orbs, the roles of individual team members, how to work with a team, and, of course, how to perform an investigation. You will learn how to review evidence in a thorough manner with an emphasis on explaining anomalies as natural occurrences. You will also learn the conceptual aspects of paranormal research, such as thinking for yourself, how to deal with Malevolent Inhuman Spirits, Human Spirits, and Poltergeists. And there is much more!

Several case studies using actual photographs, audio recordings, and videos are also incorporated for your benefit. Also included are the forms you need in the investigation process, as well as where to find basic equipment and free software used in your investigations and evidence review. This course also features an extensive and comprehensive glossary of paranormal terms to aid in your understanding of this exciting field.

This course will provide you with an excellent knowledge of the paranormal field and solid investigation techniques. But it is meant as a starting point only. Once you have completed this course, you are encouraged to go out with an open mind to learn and discover the true nature of what we commonly call, "paranormal." Your knowledge as a Paranormal Investigator will continue to grow from your own actual experiences in the field.

I am happy to have you aboard this exciting journey into the unknown! I have a lot of pride in the SPIRITeam and our approach to life and the paranormal. I believe our members are among the best in the field. That's why this course came into being; I wanted all our members to be on the same page, so to speak, and I wanted to share the best of what the Team and myself have learned over the years to help others in their own growth and professionalism. I hope you will also take great pride in being a graduate of the *Paranormal Investigator Certification Course*!

Doug Kelley, CH, CSL, CFPI
Founder of the SPIRITeam
Co-Founder of ParaNexus
Punta Gorda, FL USA
August 24, 2007

HOW TO TAKE THIS COURSE

T aking this course is fairly straightforward. Simply begin with Chapter 1, and study through the end. Some chapters contain multimedia content available in the *Online Resource Library*. Make sure you understand a chapter completely before moving on to the next. The idea is to NOT skim the information, nor skip over certain parts. This course contains no filler or fluff. ALL of the information in this course is valid to your certification as a Paranormal Investigator and has been designed for computer novices as well as those more experienced. Repetition throughout chapters is used intentionally to emphasize key points.

ONLINE RESOURCE LIBRARY

The *Online Resource Library* contains full-color pictures, videos, and video tutorials that you will need to complete this course. To access the *Online Resource Library*, open your web browser and go to:

http://cpi.ParaNexus.org

To login, use the following information:

- Username: ******* (case sensitive)

- Password: ******* (case sensitive)

If the login information changes, you will be notified via email.

Important:

- Please do not share the login information with anyone else. It is only for you and other students of this course.

- Do not save any pictures, videos, tutorials, or audios from this site to your computer. These are all copyrighted resources and are provided for your use in taking this course.

- Do not post any pictures, videos, tutorials, or audios from this site anywhere else, including discussion forums and websites.

- Do not publish or distribute any content from this site to any publication.

Your cooperation is greatly appreciated!

GETTING HELP WITH THIS COURSE

Help is available in the ParaNexus Forums at www.ParaNexus.org. A link is also available in the *Online Resource Library*. You must register for this forum. A three-month membership included free of charge with this Certification Course.

You can also email the author at dougk@spiriteam.org.

TAKING THE FINAL EXAM

Once you have completed studying the chapters and feel that you have a solid understanding of this course, the next step is the final online exam. Instructions for taking the final exam are available at:

www.ParaNexus.org/****.htm**

The final online exam will consist of multiple-choice questions, a real life investigation, and a case report. You will need a score of 90% in order to pass. You may re-take the exam until you pass it. Complete details are discussed on the Final Exam webpage.

ICONS USED IN THIS COURSE

Several icons are used in various chapters to aid in visual understanding. The table below shows the various icons used.

Scales of Scrutiny Icons

These icons give a quick visual summation as to whether specific phenomena are best explained Scientifically, Paranormally, or whether the weight of evidence is inconclusive.

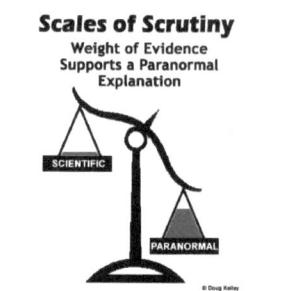

The icons below are the same as above, but in a different format.

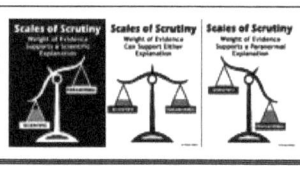

Section 1
Theories & Concepts

CHAPTER 1

Section 1: Theories & Concepts

The Calling and Mission of Paranormal Investigators

Upon Completion of this Chapter, You Will Know:	Online Resource Library
<ul><li>The SPIRITeam's Vision, Mission, and general statements.</li><li>What makes the SPIRITeam different from most groups.</li><li>Your own Vision, Mission, and general statements.</li><li>What it means to be a Paranormal Investigator.</li></ul>	There are no media files for this chapter.

S o, you want to be a Paranormal Investigator? Good for you! The field of paranormal investigation is exciting, but not necessarily new. People have been asking the same questions about the "Great Beyond" for millennia, and even with all the centuries of human wonderment and exploration, we are still not much closer to understanding the true nature of life beyond the "veil" with provable certainty. The real advantage that we have over our ancient colleagues is better technology and a more enlightened *true scientific* approach. We are seeing more and more people becoming Paranormal Investigators who refuse to be enslaved by ancient religious and mythical belief systems that really don't answer the big questions with any kind of substantive truth.

The SPIRITeam's mission is all about finding solid answers with a view to helping others along the way. We are not your typical "ghost-hunting" group. We are a professional *coterie*[1] of like-minded and serious investigators seeking to find answers in a world where few answers exist.

With this in mind, here are our Vision, Mission, and general statements:

OUR MISSION: *To discover more about ourselves, life, the human condition, the Universe, and the Grand Scheme of Things by professionally and ethically investigating paranormal events using the True Scientific Method.*

OUR VISION: *To understand and help to understand.* We endeavor to help individuals who are experiencing unexplainable events to understand *why* these events are happening through interviews, investigations, and education, and we do so without charge to the client. We endeavor to resolve the events so that all can live in peace. We seek first to help and resolve the human aspect of any paranormal incidents, and then we seek to help and resolve the

---

[1] *Coterie* means "a select group of people with a common purpose who associate frequently." —Based on definitions from www.Dictionary.com

CHAPTER 2

Section 1: Theories & Concepts

The Impact of Belief on Paranormal Investigations Part 1

The Scientific Approach vs.
the Spiritual Approach to Investigating

"A half-open mind is still half-closed."
—Doug Kelley

Upon Completion of this Chapter, You Will Know:	Online Resource Library
• The role that *Belief* plays in our approach to investigating the paranormal. • The position that religion takes toward the paranormal. • The position that mainstream science takes toward the paranormal. • The difference between a Scientific Approach and a Spiritual Approach. • The definition of a *Blended Approach*. • The definition of the *True Scientific Method*.	There are no media files for this chapter.

P aranormal investigation groups seem to be sprouting up all across the United States in recent years, no doubt due in part to increased media attention. Television shows like "Ghost Hunters," "Most Haunted," and numerous documentaries do very well because many people want to know the answer to perhaps the biggest question of humanity: What happens when we die? Like many others, I have had the same question all my life and that's why I founded the Scientific Paranormal Investigation, Research, and Identification Team (SPIRITeam) to search for the answer to that age-old and elusive question, and to help people along the way.

As I conceived of and created the SPIRITeam (initially for a novel), I wanted to approach paranormal investigations from a scientific standpoint; I wanted to use the scientific method to gain evidence of the paranormal, whichever sub-category it fell into, e.g., ghosts, UFO's, mysterious creatures, etc. That meant that I would first seek a rational, natural explanation of specific events before I would conclude that they were paranormal in nature. My philosophy would be *Open-Mindedness Tempered by Healthy Skepticism Tempered by Open-Mindedness*. Many paranormal groups also take a similar approach.

As I read about the investigations of other groups as well as performed my own investigations, I began to see a common and disconcerting theme: Many groups would have personal paranormal experiences during investigations, at times even capturing apparitions on video, but then what? How can you scientifically prove that anything paranormal happened? By definition, "paranormal" basically means something that is "beyond

CHAPTER 3

Section 1: Theories & Concepts

The Impact of Belief on Paranormal Investigations Part 2

How the Investigator's Self-Belief Impacts Paranormal Investigations

Upon Completion of this Chapter, You Will Know:	Online Resource Library
• Whether a spirit can bother or harm you. • The impact that self-belief has on paranormal investigations. • Four Universal Laws of the *Life Leadership Paragon*. • Three Universal Laws of personal responsibility. • How to have healthier relationships with everyone. • How to boost your self-concept. • How to view spirits.	There are no media files for this chapter.

When considering the paranormal field, one fact comes clearly into focus: many people, including "authorities" such as paranormal investigators, scientists, and religionists, have many differing and often conflicting theories about what is actually happening.

Some believe in demonic possession from Malevolent Inhuman Spirits (MIS); some don't. Some believe that "demons" are fallen angels; some don't. Some believe that the clergy can exorcise spirits; some don't. Some believe that simply calling on God's name and/or Jesus' name will immediately stop demonic activity; some don't.

The ironic fact of the matter is that they are all correct. How can this be? *Belief.*

I have come to believe from my research and investigation that it all boils down to one irrefutable fact: *Nobody—and I mean nobody—truly knows what in the hell is going on.* That's why we are paranormal "investigators." We want to find out.

While there is no question that *something* is going on in the world of hauntings, most paranormal events can be explained by natural occurrences. And in most cases, I believe it comes down to a person's beliefs and belief system, and this includes the various beliefs of paranormal investigators. Some believe in demons or

The Nature of Ghosts, Spirits, and Hauntings

"In the absence of positive proof, whether spirit entities exist or don't exist, thinking—and therefore belief—makes it so." –Doug Kelley

Upon Completion of this Chapter, You Will Know:	Online Resource Library
<ul><li>The nature of spirits as we understand them.</li><li>The Survival Hypothesis.</li><li>Why spirits are generally the same as us.</li><li>Common definitions of terms associated with various spirits and hauntings.</li><li>12 different types of hauntings and the investigative steps necessary to resolve them.</li><li>The meaning of PK and RSPK.</li><li>Why some spirits remain earthbound.</li><li>39 ways spirits can manifest.</li><li>The abilities spirits possess to interact with the physical realm.</li></ul>	There are no media files for this chapter.

D o ghosts and spirits exist? The answer depends on who you ask. Some people believe in ghosts and spirits, some don't. Aside from a person's personal beliefs, is there any scientific evidence to prove the existence of ghosts? The answer is no. Well, maybe.

"The Survival Hypothesis" is a theory that states a person's personality and consciousness survive the physical death of the body. Accordingly, humans have a dual nature: a physical body and an ethereal aspect that contains the personality and consciousness. This ethereal aspect has been called the soul, spirit, and "Higher Self," among others.

Many theories exist to support *The Survival Hypothesis* including the *First Law of Thermodynamics*, also known as the *Law of Conservation of Energy*, which states, "Energy can be changed from one form to another, but it cannot be created or destroyed. Thus, the total amount of energy available in the Universe is constant."

This Law is supported by Albert Einstein's famous equation, $E = mc^2$, which describes the relationship between energy and matter. In this equation, known as the *Theory of Relativity*, energy (E) is equal to mass (m) times the square of a constant (c). Thus, Einstein suggested that energy and matter are interchangeable. His equation also suggests that the *quantity of energy and matter in the Universe is fixed.*

What does this mean to us as human beings trying desperately to understand what happens to us at death? If true, it means that we do not die, but are simply "converted" from one form of energy into another—the

CHAPTER 5

Section 1: Theories & Concepts

The Impact of Belief on Paranormal Investigations Part 3

The Psychology of Paranormal Investigations

Upon Completion of this Chapter, You Will Know:	Online Resource Library
<ul><li>Your dominant communication and behavioral style.</li><li>The difference between Assertive, Aggressive, and Passive behavior.</li><li>The nature and meaning of the "energy" we all emit.</li><li>The meaning of "boundaries," and how to establish healthy ones.</li><li>The significance of boundaries when performing paranormal investigations.</li><li>How to deal effectively with benevolent and malevolent spirits.</li><li>The meaning of "possession" and how to prevent it.</li><li>How to handle chronic spirit attacks.</li><li>How to think for yourself.</li></ul>	There are no media files for this chapter.

S o far in this course, we have discussed at length the critical impact that *Belief* has on your role as a serious Paranormal Investigator. This includes self-belief as well as the beliefs you hold about life, the paranormal, and everything else. We also discussed at length the nature of spirits and the most common types of hauntings. We're going to add a little more foundation to what we've already learned and in the process, bring it all together to explain exactly why your beliefs will affect the success or failure of an investigation.

To recap, in order to have healthy relationships and healthy interaction with others, you must accept and respect yourself (Self-Acceptance), take responsibility for your actions (Self-Responsibility), and respect the rights of others (Respect for Free Will). These three skills are actually Universal Laws and when combined, they create the synergy of Self-Completeness. Self-Complete people know who they are, and why they are who they are. They recognize that as adults, they now have the ability to choose their own path and can no longer use the excuse, "But that's how I was raised," which is no different than saying, "I was just following orders." Self-Complete people take responsibility for who they are, what they think, what they believe, and how they deal with others.

CHAPTER 6

Section 1: Theories & Concepts

The Nature of Orbs

Upon Completion of this Chapter, You Will Know:	Online Resource Library
• Common beliefs about orbs. • The scientific explanation of orbs. • 4 critical factors in photography. • How to identify various types of orbs. • Methods of reducing orbs in photographs. • Criteria for determining true orbs. • How belief impacts orb phenomena.	*Chapter 6 Media* http://CPI.ParaNexus.org Username: ******* Password: *******

O rbs are a highly, highly controversial subject, to say the least. On one end of the spectrum, many people, including some paranormal researchers and ghost hunters, believe that some orbs are genuine evidence of spirit activity. On the other end of the spectrum, you have a growing number of paranormal investigators who believe that most orbs are usually nothing more than airborne particles such as dust. And of course, many people fit somewhere in the middle. Orbs are also referred to as plasmoids, balls of light, light spots, Basic Life Forms (BSF), and spheres.

The orb phenomenon is really a modern day enigma. Since the late 1990's, many people have been capturing strange anomalies on their cameras, and not surprisingly, the vast majority of these picture anomalies have been captured using consumer compact digital cameras. Although digital cameras have been in development since the early 1970's, they really didn't hit the mass market until the mid to late 1990's.

Since that time, digital cameras have literally exploded in the marketplace. Interestingly, this is also when the phenomena involving orbs, light streaks (energy rods), and vortexes really became so widespread. As consumer compact digital cameras get smaller in size, the reports of orb phenomena increase due to the short distance of the flash to the lens.

Before the advent of digital cameras, 35mm cameras, Kodak's 126 and 110 Pocket Instamatics, Polaroid, and others were the norm, and orbs were certainly not a common phenomenon. Even today, current 35mm cameras yield almost zero orbs, and Digital SLRs also greatly minimize them as well.

CHAPTER 7

Section 1: Theories & Concepts

Other Visual Anomalies

Upon Completion of this Chapter, You Will Know:	Online Resource Library
• How Mists are formed and how to identify naturally occurring mists with examples. • How to identify a potentially paranormal ecto-mist. • The true nature of Energy Rods and how to analyze photographs for such. • The nature of Vortexes and how to avoid creating the effect in your pictures. • The cause of lens flare. • About ethereal faces. • About infrared anomalies and what to watch out for. • The uncertain nature of Ghost Lights.	*Chapter 7 Media* http://CPI.ParaNexus.org Username: ****** Password: ******

N ow that our discussion of orbs is out of the way (whew!), let's examine several other anomalies you will experience from time to time on your investigations, such as mists, vortexes, lens flare, and others. Again, our first approach is to explain these phenomena in a scientific manner.

ECTO-MISTS

Scales of Scrutiny
Weight of Evidence Can Support Either Explanation

SCIENTIFIC — PARANORMAL

© Doug Kelley

Mists have been reported for centuries, and many photos exist showing the unusual phenomena. Some call mists "ectoplasm mists," "ecto-mists," or just "ecto," which, as we've already discussed, is more or less a modern term for an early twentieth century staged phenomenon among less than honest Mediums. We use it loosely today to refer to mists or other substances seemingly left behind by spirit entities. Ecto-mists usually appear as white or pale gray, but black mists have also been reported, which are also called Shadow People.

Ecto-mists show up in photos occasionally and often confound the photographer, especially if he or she is not a Paranormal Investigator. Clients who submit mist pictures always claim that no mist was present when the picture was taken. In this instance,

EVPs and Spirit Communication

Upon Completion of this Chapter, You Will Know:	Online Resource Library
• The nature of Electronic Voice Phenomena and its impact on paranormal research. • How to properly conduct an EVP vigil. • The nature of séances. • The nature of table tipping and how to participate. • The true nature of Ouija Boards. • How belief about séances impacts effective paranormal investigations.	*Chapter 8 Media* http://CPI.ParaNexus.org Username: ****** Password: ******

S pirit communication is and has always been among the most controversial topics in the world, and the enigmatic human behavior concerning it confounds those who study psychology and teach human development—and I'm one of them. When you think about it, the very foundation of the world's religions is based on spirit communication. And here is where the enigmatic part comes in: the very person who would condemn another for getting a psychic reading turns right around, goes to church, and performs "spirit communication" with whatever god they worship. The same holds true for those who burned witches and free thinkers at the stake during the dark ages on Saturday night, and then went to church the next morning to "commune" with the "Great Spirit."

Every time a person prays, he or she is communicating with a spirit. "But," you say, "that spirit is God!" Yes, but one person's god is not necessarily the same as another person's god, and each religion pretty much forbids communicating with or calling on some other religion's god. Plus, some religions pray to more than one spirit, and other religions are polytheistic.

The Christian religion was literally founded on spirit communication, and the Bible is filled with it. "Prophet" is just another label for Psychic, or Medium, or Sensitive, or Intuitive—it's just an acceptable label that fits a certain belief system. Prophets were spirit communicators and communicated with spirits other than God; angels were included as well.

CHAPTER 9
Section 1: Theories & Concepts

Working with Psychics, Mediums, Sensitives, and Intuitives

Upon Completion of this Chapter, You Will Know:	Online Resource Library
<ul><li>The various types of psychic abilities.</li><li>Preferred titles of those with psychic abilities.</li><li>The human side of Psychics.</li><li>How to determine the authenticity of a Psychic.</li><li>How to recognize Cold Reading and Hot Reading.</li><li>How to work with Psychics on an investigation.</li></ul>	There are no media files for this chapter.

The most critical aspect of a blended approach to paranormal investigations is the availability of Sensitive team members. In my opinion, not having them on the team removes a significant and valuable tool for finding answers to the questions we all seek. If you are Sensitive yourself, then you understand what I mean. If not, then following are some helpful tips in working with Sensitives.

Different Sensitives have different abilities, and to varying degrees. Following are a few of the common types of psychic gifts. All fall under the general name of Extrasensory Perception (ESP). Please see the Glossary for more information on each one.

- Clairvoyance (seeing)
    - Precognition (seeing future events)
    - Retrocognition (seeing into the past)
- Clairsentience / Psychometry (feeling/touching)
- Clairaudience (hearing/listening)
- Claircognizance (knowing)
- Telepathy (mind reading)
- Mediumship / Channeler
- Empathic (feeling the emotions of others)

Other less common terms also exist to identify varying psychic abilities.

Working With and Contributing to a Team

Upon Completion of this Chapter, You Will Know:	Online Resource Library
• The *Two Most Important Relationship Questions*. • How to effectively work with a team. • The importance of excellent communication skills. • Why a positive attitude is critical to paranormal investigating. • Methods of contributing to the team and the paranormal field. • How to view other paranormal research groups. • How to view and handle critics, hard-core skeptics, and debunkers.	There are no media files for this chapter.

eing a Paranormal Investigator is a noble endeavor as it shows that you are not satisfied with the status quo of accepting the world on other people's terms and with other people's beliefs. Another advantage of being a Paranormal Investigator is that you can still explore the unknown even if you can't travel the world in search of its mysteries.

A PRIMER IN HEALTHY RELATIONSHIPS

In chapter 5, I discussed how essential Self-Acceptance, Self-Responsibility, Respect for Free Will, and Self-Completeness are to healthy relationships—especially the relationship you have with you. Another essential ingredient to healthy relationships is what I call, *The Two Most Important Relationship Questions*. I personally believe these two questions alone are worth the price of this course thousands of times over. Not only will these two questions help you with team relationships, they will help you to make *all* of your present and future relationships healthy.

Again, if you are wondering what this has to do with being a Paranormal Investigator, it has *everything* to do with it. Not only do you have relationships with clients, you also have relationships with your teammates.

CHAPTER 11

Section 1: Theories & Concepts

Roles of Investigators within a Team

Upon Completion of this Chapter, You Will Know:	Online Resource Library
• The various team member roles and positions. • How to perform historical research. • Sources for historical research. • The organizational structure of a team.	There are no media files for this chapter.

A s a team member or prospective team member, it is important that you understand the respective roles each member plays to ensure a successful group and investigation. All members can contribute value to the group, as discussed in the previous chapter. As you spend time with the team, your role could change, therefore, you should be familiar with the various roles of the team. The following roles are suggested, but are subject to your own group's Standards and Protocols.

CORE MEMBER POSITIONS

The primary members who participate in all investigations are the Founder/Director, Equipment Manager, Case Manager, and a Sensitive member. Other members will be invited as the specific case requires and allows. High profile cases as well as larger locations will include the team Documentarian as part of the core group. Larger locations will of course accommodate many team members, but private residences will accommodate only a few.

CLEANSING TASK FORCE

This sub-group of the team consists of established and experienced members who have been trained in performing cleansings, and possess great strength of mind, emotions, and belief. Hopefully, this describes every member. When a cleansing is anticipated on an investigation, these members will be invited to participate

Section 2
Tools & Equipment

CHAPTER 12

Section 2: Tools & Equipment

Cameras and Audio Recorders

Upon Completion of this Chapter, You Will Know:	Online Resource Library
<ul><li>The different types and benefits of cameras.</li><li>The potential uses and benefits of Infrared Photography.</li><li>About video cameras and camcorders.</li><li>Photography basics.</li><li>About Digital Video Recorder systems.</li><li>How Thermal Imagers can aid in your research (if you can afford one).</li><li>About audio recorders.</li><li>Digital vs. analog recorders.</li></ul>	*Chapter 12 Media* http://CPI.ParaNexus.org Username: ****** Password: ******

Since the scientific side of the blended investigation approach requires electronic equipment to complete, we will discuss the most common tools that you will use over the next three chapters. This chapter will review still cameras, video cameras, and audio recorders. Serious Paranormal Investigators will want to become proficient at operating basic equipment. Once you've done it a few times, it gets easy.

DIGITAL AND FILM CAMERAS

A lot of debate exists over the issue of digital verses film cameras for paranormal investigations. The pro-film camp says that 35mm cameras don't pick up dust orbs, which is true for the most part. The digital camp says that with the right camera, dust orbs are not an issue, which is also true. This being the case, personal preference is usually the deciding factor.

From our discussion on orbs, the key to minimizing them is to use a camera that doesn't have the flash so close to the lens. This usually means getting an SLR type camera, as opposed to the consumer compact style. In my opinion, the features of digital cameras far outweigh the issue of which one is the best choice. Ideally, the evidence I'm looking for needs to be better than just an interesting still shot that has no orbs. My suggestion

Electro-Magnetic Fields and EMF Meters

Upon Completion of this Chapter, You Will Know:	Online Resource Library
<ul><li>The types of Electro-Magnetic Field (EMF) meters available.</li><li>The nature of electro-magnetic fields and how they are generated.</li><li>Types of electro-magnetic fields.</li><li>How EMF meters work.</li><li>How to use an EMF meter during an investigation.</li><li>The potential effects of electro-magnetic fields on humans and how this can explain some paranormal activity.</li></ul>	There are no media files for this chapter.

Electro-Magnetic Field (EMF) meters can be a significant tool among the comparatively few available to paranormal investigators. EMF meters detect the *fluctuation* of magnetic fields, electric fields, radio frequencies, and microwave energy, depending on the model. Although not proven, empirical evidence suggests that spirits can affect electro-magnetic fields, which can then be detected by EMF meters. Since the veracity of EMF meters has not been proven, any advertising with claims that an EMF meter has "ghost detection" capabilities is purely marketing hype.

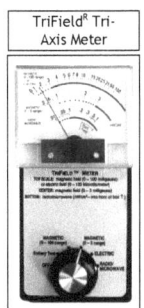

TriField® Tri-Axis Meter

Many types of EMF meters are available and generally fall into one of two categories: single axis and tri-axis. Single axis meters measure only one dimension of the electro-magnetic field, in other words, you have to tilt and turn the meter in three directions to get a full measurement or to locate the source. A tri-axis meter measures all three axes simultaneously. Tri-axis meters are usually more expensive, however, AlphaLabs makes a model of the TriField® meter that runs around $130. Some common models used in paranormal research are shown below.

Many of the most common EMF meters used in paranormal research detect AC (alternating current) fields; but some, such as the TriField® Natural EM Meter, ignore AC and only detect DC (direct current) fields. The overwhelming majority of AC fields are man-made and include electrical wiring, appliances, microwave ovens, etc. Anything that is plugged

Thermometers and Other Investigation Tools

Upon Completion of this Chapter, You Will Know:	Online Resource Library
<ul><li>About other tools for research including:<ul><li>Infrared and ambient thermometers</li><li>Communication equipment</li><li>Flashlights</li><li>IR extenders / illuminators</li><li>Parabolic microphones</li><li>Compasses</li><li>Motion detectors</li><li>Geiger Counters</li><li>Ion generators/detectors</li><li>Ghost Box</li></ul></li><li>The nature of Trigger Objects and how they are used.</li><li>Subjective tools including dowsing rods and pendulums.</li><li>Basic equipment list for every investigator.</li></ul>	There are no media files for this chapter.

T his chapter lists a variety of the common tools available for paranormal investigations. Some you will use more often; some you will use only when the circumstances require it. Many Paranormal Investigators create and build custom designed tools for special use. Just like a jigsaw puzzle, each tool can provide a part of the overall picture to give a better understanding of what might be occurring on a specific investigation.

THERMOMETERS

Thermometers are used in paranormal investigations for registering any cold or hot spots reportedly associated with spirit activity. Non-Contact Infrared thermometers are perhaps the most common type used by paranormal investigators. The advantage of this type of thermometer is that you can "point and shoot" to get an instant reading of the surface temperature of a room or an object. Holding down the trigger will allow you to take continuous readings. Most models also utilize a laser to pinpoint exactly where you are taking the reading, as well as a backlit display for easy viewing in the dark.

Section 3
Performing an Investigation

How to Perform a Paranormal Investigation

Upon Completion of this Chapter, You Will Know:	Online Resource Library
<ul><li>What to do before an investigation.</li><li>How to handle first contact with clients.</li><li>What cases are acceptable to investigate.</li><li>Where to find investigation locations.</li><li>How to prepare for an investigation.</li><li>The best times to investigate.</li><li>What to take on an investigation.</li><li>How to make your approach professional.</li><li>How to prepare a location.</li><li>What to do during the investigation.</li><li>Tips on equipment setup and client interviews.</li><li>Safety issues to be concerned with.</li><li>How to conduct a vigil.</li><li>How to perform Baseline Tests.</li><li>Protocols for successful investigations.</li></ul>	There are no media files for this chapter.

This chapter contains numerous details that must be followed for a successful paranormal investigation. Please refer to the individual chapters in this course for detailed discussions of respective points. It is imperative that you become extremely familiar with every detail so the entire investigation runs smoothly. A paranormal research team who knows their "stuff" makes a good and credible impression on clients, not to mention reassuring them that you can actually help them. Professional appearance and demeanor will go a long way in helping the client—especially in view of the fact that most clients only call paranormal investigators as a last resort.

This chapter was written in such a way so as to accommodate a team working together or an individual working with a partner. If working with a team, bear in mind that not all members will necessarily participate in every investigation. Core members will usually always participate, and other members will be invited as

How to Analyze and Review Evidence

Upon Completion of this Chapter, You Will Know:	Online Resource Library
<ul><li>How to organize your computer folders for case and evidence files.</li><li>How to prepare picture, audio, and video files for uploading to the member's area website.</li><li>How to name your files for easy recognition.</li><li>How to pace yourself when reviewing evidence.</li><li>Natural explanations vs. paranormal explanations.</li><li>Proof of paranormal activity.</li><li>How to analyze audio files for EVPs (with case studies).</li><li>The classes of EVPs.</li><li>How to analyze photos (with case studies).</li><li>The meaning and use of Exif information.</li><li>How to analyze video files (with a case study).</li><li>EMF and temperature analysis.</li><li>What to include in a case report.</li><li>How to store evidence after the investigation.</li></ul>	*Chapter 16 Media* http://CPI.ParaNexus.org Username: ****** Password: ******* Video Tutorials Included in this Chapter: <ul><li>*How to ZIP Image Files*</li><li>*How to Use an Audio Editor*</li><li>*How to Use a Photo Editor*</li></ul>Software Included in this Chapter: <ul><li>*EasyZip*</li><li>*Audacity*</li><li>*IrfanView*</li></ul>

 vidence analysis and review is one of the most important parts of being a Paranormal Investigator—if not the most tedious. Evidence review is the part of the investigation that separates the serious-minded from the "Looky Loos" who are only interested in the "thrill of the hunt."

Evidence review can actually be exciting and fulfilling when you find an anomaly that requires you to put on your analytical hat and search hard to find an answer. Evidence review is one of the primary methods of gaining experience in the paranormal field, and gaining experience is one of the reasons why *every* team member should help with evidence review. The burden of reviewing evidence should never fall on one or two people among the team—especially when several team members participated in the investigation.

To make it possible for everyone to help with evidence review, all participating members should upload their pictures and audios to the member's area as soon as possible after the investigation.

Section 4
Revealing, Explaining, Resolving, & Cleansing

Revealing Your Evidence: How to Help Clients and Spirits

Upon Completion of this Chapter, You Will Know:	Online Resource Library
 • How to do a Reveal. • How to help the client. • How to help the spirit. • What to say to help the spirit move on.	There are no media files for this chapter.

 t this point, you have performed your investigation, reviewed the evidence, and are now ready to inform your client as to your findings. You may or may not have found any paranormal activity, which is fine. You may have to do a follow up investigation if the circumstances warrant.

HELPING THE CLIENT

One of the primary purposes of your investigation is to resolve the client's issues with the activity, and just by you giving him or her a kind ear will do much good in resolving the situation. The most common thing we hear from clients is that they are worried that others will think they're crazy. By simply listening to and investigating their concerns, you are doing much good.

If you were not able to find any paranormal explanations, reassure the client that whatever was happening wasn't paranormal in nature. If you did find some strong evidence of paranormal activity, you should also share that information, but diplomatically.

When revealing paranormal results, choose your words wisely so that you don't unnecessarily frighten the client without cause. And of course, there is rarely cause to be overly concerned with spirit activity unless it is causing problems for the client. Make sure that your disposition and demeanor is friendly, personable, and positive. Smile!

CHAPTER 18

Section 4: Revealing, Explaining, Resolving, & Cleansing

How to Cleanse Malevolent Human and Inhuman Spirits

The person who is self-possessed will not become possessed by anyone else—human or spirit.
— Doug Kelley

Upon Completion of this Chapter, You Will Know:	Online Resource Library
<ul><li>The prevalence of disabling beliefs about malevolent spirits.</li><li>Whether malevolent spirits can attack or harm you.</li><li>The psychology of cleansing malevolent spirits.</li><li>Confronting a malevolent spirit.</li><li>About exorcisms.</li><li>Cleansing aids and why they work.</li><li>What to do before cleansing malevolent spirits.</li><li>How to smudge a home or area.</li><li>The *Energy Burst Protocol* cleansing ceremony.</li><li>A syllabus of a cleansing ceremony.</li><li>The ultimate key to successful cleansings.</li></ul>	*Chapter 18 Media* http://CPI.ParaNexus.org Username: ****** Password: ****** Video Included in this Chapter: • *A Live Cleansing Ceremony*

O n rare occasions, you may discover that you are dealing with a malevolent spirit on an investigation. This spirit will usually be an Inhuman Spirit, but Human Spirits can also get pissy at times. As discussed in Chapter 5, *The Psychology of Paranormal Investigations*, there is an amazing amount of information online concerning Malevolent Inhuman Spirits—much of it disinformation based on preconceived notions, religious superstition, and disabling beliefs. As a reminder of just how deep these mass beliefs run, consider the following partial article concerning Malevolent Inhuman Spirits. By the way, it was written by a paranormal research group and is presented exactly as found:

The Malevolent Inhuman Spirit is a being of great energy which is only matched by its often terrifying intelligence. Do not think that you have an upper hand or have outmatched them at

The Art of Being a Paranormal Investigator

e have examined much throughout this course. If you have studied and absorbed it well, you have gained an excellent knowledge of the paranormal field, solid investigation techniques, and excellent evidence review capabilities. You have learned all about the many types of hauntings, how to view spirits, how to work with Sensitives, how to use equipment, and much more.

However, it is my sincerest hope that you have gained far more than just head knowledge. I hope you have gained new insight into yourself as a Real Live Human Being. Regardless of what you want to do in life, it all begins and ends with how you view yourself. To be successful at any endeavor you decide to undertake, you must have Self-Acceptance, Self-Responsibility, Respect for Free Will, and Self-Completeness. These qualities are also indispensable in the field of paranormal investigation.

You also learned the critical importance of *Belief*, and how it impacts everything you do—especially your approach to paranormal research. *Belief* also has an enormous impact on the client and is often the very reason that he or she is experiencing paranormal activity. To this end, it is very important to coach the client on every investigation.

The art of being a Paranormal Investigator is all about looking beyond what you see and trying to discern new insights into life, existence, and the Universe. It is about knowing who you are, and *why* you are who you are. It is about being yourself and sharing the gift of yourself with all those you meet. It is about lightening the load of others, because when you lighten their load, you also lighten your own.

Being a Paranormal Investigator is a noble calling, and therefore, it is about contributing to the field and the team in meaningful ways. It is about not taking yourself too seriously; being able to laugh at yourself, and treating others with respect and dignity—including spirits.

I hope you will take the knowledge you have gained from this course, and continually add to it by gaining experience in the field. Go out and make your mark; discover new things, and stand tall with your head held high! Be among the best there is, and take a great deal of pride in your new status as a Certified Paranormal Investigator!

Namastê!

HOW TO TAKE THE FINAL EXAM

Once you have completed studying the chapters and feel that you have a solid understanding of this course, the next step is the final online exam. Instructions for taking the final exam are available at:

www.ParaNexus.org/******.htm

The final online exam will consist of multiple-choice questions, a real life investigation, and a case report. You will need a score of 90% in order to pass. You may re-take the exam until you pass it. Complete details are discussed on the Final Exam webpage.

APPENDIX

APPENDICES

Enroll in the CPI Course Here: http://www.paranexus.org/forum/index.php?page=51

Course is available with or without certification